YOUR KNOWLEDGE HAS VALUE

- We will publish your bachelor's and
  master's thesis, essays and papers

- Your own eBook and book -
  sold worldwide in all relevant shops

- Earn money with each sale

Upload your text at www.GRIN.com
and publish for free

Imprint:

Copyright © 2012 GRIN Verlag, Open Publishing GmbH
Print and binding: Books on Demand GmbH, Norderstedt Germany
ISBN: 9783668253339

This book at GRIN:

http://www.grin.com/en/e-book/335107/the-hesse-pencil-and-its-cayleyan

Jakob Bongartz

Aus der Reihe: e-fellows.net stipendiaten-wissen

e-fellows.net (Hrsg.)

Band 1859

The Hesse pencil and its Cayleyan

GRIN Publishing

GRIN - Your knowledge has value

Since its foundation in 1998, GRIN has specialized in publishing academic texts by students, college teachers and other academics as e-book and printed book. The website www.grin.com is an ideal platform for presenting term papers, final papers, scientific essays, dissertations and specialist books.

The Hesse pencil and its Cayleyan

Jakob Bongartz

23. Juli 2012

Bachelorarbeit Mathematik

Mathematisches Institut

Mathematisch-Naturwissenschaftliche Fakultät der

Rheinischen Friedrich-Wilhelms-Universität Bonn

Contents

Einleitung

Wir wollen eine kurze Zusammenfassung des Inhaltes der Bachelorarbeit geben. Im ersten Teil behandeln wir die Konfiguration der neun Wendepunkte einer elliptischen Kurve im $\mathbb{P}^2$. Betrachtet man neben den Wendepunkten noch die zwölf Geraden durch je zwei der Wendepunkte, so ergibt sich die sogenannte *Hessesche Konfiguration* $(9_4, 12_3)$. Dieses Symbol bedeutet, dass man neun Punkte und zwölf Geraden hat, derart dass jeder Punkt auf vier der Geraden liegt und jede Gerade drei der Punkte enthält. Diese Konfiguration wurde schon früh von Colin Maclaurin (1698-1746) entdeckt und später im Detail von Otto Hesse (1811-1874) studiert, nach dem sie dann auch benannt wurde. Sie ist eindeutig bis auf Isomorphie bestimmt, kann aber nicht in beliebigen Räumen realisiert werden, z.B. nicht in der affinen Ebene über $\mathbb{R}$. Die einfachste Realisierung bieten die neun Punkte und zwölf Geraden in der affinen Ebene über $\mathbb{F}_3$. Wir werden die *Hesseschen Konfiguration* auf zwei verschiedene Arten als Realisierung der Wendepunkte elliptischer Kurven sehen.

Die erste ist durch das *Hessesche Büschel* induziert, welches ein Linearsystem von kubischen Kurven in $\mathbb{P}^2$ darstellt, gegeben durch

$$E_{t_0, t_1} : t_0 (Z_0^3 + Z_1^3 + Z_2^3) + t_1 Z_0 Z_1 Z_2 = 0,$$

wobei $[t_0 : t_1] \in \mathbb{P}^1$. Das Büschel enthält vier singuläre Kurven. Diese sind jeweils die Vereinigung von drei verschiedenen Geraden, die Geraden der Konfiguration bilden. Die Punkte werden durch die Basispunkte des Linearsystems gegeben, welche den Wendepunkten aller glatten Elemente des Büschels entsprechen. Wir werden sehen, dass diese Punkte und Geraden vom Typ $(9_4, 12_3)$ sind. Weiterhin werden wir zeigen, dass das *Hessesche Büschel* eine nicht eindeutige Normalform für elliptische Kurven induziert. Auch werden wir eine *Weierstrass Normalform* bestimmen, um dann für glatte Kurven aus dem *Hesseschen Büschel* die j-Invariante zu bestimmen.

Die zweite Realisierung bedient sich keiner konkreten Kurven sondern der Gruppenstruktur auf allgemeinen elliptischen Kurven. In dieser Gruppe stimmen die 3-Torsionspunkte mit den Wendepunkten überein. Die Geraden werden gegeben durch die Menge der Geraden durch jeweils drei der Wendepunkte. Wir werden außerdem sehen, dass die Wendepunkte eine Gruppe bilden und abstrakt isomorph zu $(\mathbb{Z}/3\mathbb{Z})^2$ sind. Diese Tatsache liefert eine natürliche Verbindung zu der Realisierung der Konfiguration in $\mathbb{F}_3^2$.

Die Gruppe der projektiven Kollineationen, die das *Hesseschen Büschel* in sich abbilden, wird die *Hessesche Gruppe* G_{216} genannt. Wir werden sehen, dass diese auf natürliche Art und Weise als Untergruppe der affinen linearen Gruppe von $\mathbb{F}_3^2$ gesehen werden kann. Weiterhin werden wir beobachten, dass die Orbiten unter dieser natürlich induzierten Gruppenoperation den elliptischen Kurven mit gleicher j-Invariante entsprechen.

2

Genauer werden wir sogar zeigen, dass es genügt, die Orbiten der Tetraedergruppe $\mathfrak{A}_4$ zu betrachten, die als Faktorgruppe der G_{216} auftritt.

In dem zweiten Teil der Bachelorarbeit werden wir die Theorie der polaren Hyperflächen studieren. Mit deren Hilfe werden wir eine duale *Hessesche Konfiguration* in der dualen Ebene $\check{\mathbb{P}}^2$ definieren. Die Theorie der Polaren wird uns dabei einige äquivalente Beschreibungen der Hesseschen Kurve $He(E)$ einer ebenen Kurve E liefern. Mit Hilfe dieser Theorie werden wir zu einer Kurve E in $\mathbb{P}^2$ eine Kurve in $\check{\mathbb{P}}^2$ assoziieren, die *Cayley Kurve $Ca(E)$*. Wenn wir ein Element $E_{1,\lambda}$ des *Hesseschen Büschels* fixieren, werden wir zeigen, dass $Ca(E_{1,\lambda})$ ein Element des dualen Büschels ist. Wir werden außerdem imstande sein, die 2-Torsionspunkte einer glatten Kubik $E_{1,\lambda}$ genau zu beschreiben, nämlich als die Schnittpunkte der Kurve mit der sogenannten *harmonischen Polaren*.

Abschliessend möchte ich mich bei Herrn Prof. Dr. Daniel Huybrechts für das interessante Thema und die gute Betreuung bedanken und bei Herrn Priv. -Doz. Dr. Christian Liedtke für zahlreiche Gespräche und nützliche Ratschläge zur Gestaltung meiner Arbeit.

Introduction

In the first part of this bachelor thesis we will discuss the configuration of the nine inflection points of an elliptic curve, considered in the projective plane $\mathbb{P}_2$ over an arbitrary algebraically closed field k, and the twelve lines through three of the inflection points. This will give the famous *Hesse configuration* $(9_4, 12_3)$ of nine points and twelve lines, each point lying in four lines and each line containing three points. First, it was introduced by Colin Maclaurin (1698-1746) and later studied by Otto Hesse (1811-1874). This configuration is even unique up to isomorphism, but it cannot be realized in arbitrary planes, e.g. the affine plane over $\mathbb{R}$. A very natural realization is obtained by the nine points and twelve lines in the affine plane $\mathbb{F}_3^2$. We want to give two approaches to the study of the configuration of the inflection points of an elliptic curve.

The first one is induced by the *Hesse pencil*, a linear system of cubic curves in $\mathbb{P}^2$, given by

$$E_{t_0, t_1} : t_0(Z_0^3 + Z_1^3 + Z_2^3) + t_1 Z_0 Z_1 Z_2 = 0,$$

with $[t_0 : t_1] \in \mathbb{P}^1$. The pencil has four singular members, each one the union of three distinct lines. This gives the lines of the configuration. The points are realized by means of the nine base points of the pencil, since each one is contained in four of the lines and each line contains three of the base points. The *Hesse pencil* also induces a normal form of elliptic curves, since every smooth cubic curve is projectively equivalent to a member of the pencil. We will compute also the usual invariants, e.g. the j-invariant, by means of the computation of a *Weierstrass normal form* of a smooth member of the pencil.

The second approach is an abstract one. One defines a group law on an elliptic curve, in which the 3-torsion points coincide with the inflection points, which give the points of the configuration. The lines are defined by the set of lines through three of the inflection points. Since the group of 3-torsion points is isomorphic to $(\mathbb{Z}/3\mathbb{Z})^2$, this gives a natural connection with the realization of the *Hesse configuration* in $\mathbb{F}_3^2$.

The group of projective collineations which preserves the *Hesse pencil* is called the *Hessian group* G_{216} of order 216, which can be naturally seen as a subgroup of the affine linear group of $\mathbb{F}_3^2$. We will see, that the orbits under this group will correspond to the set of curves with the same j-invariant. More precisely, we will show that it suffices to consider the orbits of the tedrahedral group $\mathfrak{A}_4$.

In the second part of this thesis we will study the well known theory of polar hypersurfaces. By means of those we will be able to define a corresponding dual *Hesse configuration* in the dual plane $\check{\mathbb{P}}^2$. The theory of polars will give us useful equivalent definitions of the *Hessian curve* $He(E)$ of a plane cubic curve E. This will lead us to the construction of the *Cayleyan curve* $Ca(E)$ of an elliptic curve. For $E_{1,\lambda}$ in the *Hesse pencil* we find, that $Ca(E_{1,\lambda})$ is a member of the dual *Hesse pencil*. Another interesting result in this part is an explicit characterization of the 2-torsion points of the abelian group defined by an elliptic curve. We will see that these correspond to the ramification points of the projection from the inflection point, which forms the origin in the group law.

Finally I would like to thank Prof. Dr. Daniel Huybrechts for the interesting topic and the good support and I also want to thank Priv. -Doz. Dr. Christian Liedtke for numerous conversations and good advice for the arrangement of the thesis.

1 Hesse pencil and Hesse configuration

In this section, we want to consider the well-known *Hesse pencil*, which goes with the *Hesse configuration*. We will consider the *Hesse configuration* from two points of view. First, we will start with some geometrical properties to illustrate the *Hesse pencil*. Later, we will define a commutative group law on an elliptic curve and study the *Hesse configuration* in this way. We let k be an algebraically closed field of characteristic different from three, e.g. $k = \mathbb{C}$. The projective plane over k is denoted by $\mathbb{P}^2$, as usual. We denote by E an arbitrary plane cubic curve defined by the zero locus $V(F)$, where F is a homogeneous polynomial of degree three. Furthermore, we denote by $\mathbb{T}_x(E)$ the tangent space of the point $x \in E$ with respect to E. It is given by $V(\sum_{i=0}^{2} \partial_i F(x) Z_i)$ for homogeneous coordinates $[Z_0 : Z_1 : Z_2]$.

1.1 A geometrical approach

Let us start with some definitions :

Definition 1.1.1. *The* Hesse pencil *is the one-dimensional linear system of plane cubic curves in $\mathbb{P}^2$ given by*

$$E_{t_0, t_1} : t_0(Z_0^3 + Z_1^3 + Z_2^3) + t_1 Z_0 Z_1 Z_2 = 0,$$

with $[t_0 : t_1] \in \mathbb{P}^1$. We denote by $\mathcal{P}$ the set of members of the Hesse pencil, namely $\mathcal{P} := \{E_{1,\lambda} : \lambda \in k\} \cup \{E_{0,1}\}$.

In the following we use an affine parameter $\lambda := \frac{t_1}{t_0}$ and denote $E_{1,\lambda}$ by E_λ. The curve $E_{0,1}$ is denoted by E_∞.

Definition 1.1.2. *Let E be a plane cubic curve defined by $F(Z_0, Z_1, Z_2) = 0$. Then $He(E)$, the* Hessian curve *of E (or just the* Hessian*), is the plane cubic curve defined to be $V(\det((\partial_i \partial_j F)_{i,j=1,2}))$, namely the zero locus of the determinant of the Hesse matrix of F.*

A very important role in this first part of the thesis plays the following

Lemma 1.1.3. *Let E be a smooth cubic. Then the set of inflection points of E coincides with the intersection of E and $He(E)$.*

Proof. We refer to Proposition 2.5 in [Kuw11]. The proof uses the theory of polar hypersurfaces, which is discussed in Chapter 2.1 of this thesis. Note that we do not use this lemma to prove anything in the second chapter, hence the logic is not disturbed. $\qquad\square$

Remark 1.1.4. *By* Bézout's Theorem *(cf. [Ful08, Chap. 5.3], the intersection of E and $He(E)$ contains at most nine points. In fact, we will see in Lemma 1.1.10 that we have exactly nine inflection points for any smooth cubic curve.*

The importance of the Hesse pencil stems from the following

Theorem 1.1.5. *Every smooth cubic E is projectively equivalent to a member of the Hesse pencil, i.e. it exists $\lambda \in k$ with $E \cong E_\lambda$.*

Proof. We fill out the details of the proof of Theorem 4 in [BK86]. Let $p_1 \neq p_2 \in E$ be different inflection points with corresponding inflection tangent lines $\mathbb{T}_{p_i}(E)$. Choose $q_i \in \mathbb{T}_{p_i}(E) \smallsetminus (\mathbb{T}_{p_1}(E) \cap \mathbb{T}_{p_2}(E))$, $i = 1, 2$, two different points. Since p_1 and p_2 are inflection points, we have $\mathbb{T}_{p_i}(E) \cap E = \{p_i\}$. By assumption, the line $\overline{p_1, p_2}$ contains neither q_1 nor q_2. Suppose, we have a line through p_i, $i = 1, 2$, and q_1 and q_2. Then we get $q_2 \in \mathbb{T}_{p_1}(E) \cap \mathbb{T}_{p_2}(E)$ resp. $q_1 \in \mathbb{T}_{p_1}(E) \cap \mathbb{T}_{p_2}(E)$. This is a contradiction to the choice of the q_i and hence it is impossible to have three of these four points in a line, which means that they are in general position.

Since $PGL_3(k)$ acts transitively on four points in general position, we can choose coordinates $[Z_0 : Z_1 : Z_2]$ such that $p_1 := [0 : 0 : 1]$ and $p_2 := [0 : 1 : 0]$ are inflection points with corresponding inflection tangent lines $\mathbb{T}_{p_i}(E) = V(Z_i)$, $i = 1, 2$. Let E be defined by

$$F = \sum_{\nu_1 + \nu_2 + \nu_3 = 3} a_\nu Z_0^{\nu_0} Z_1^{\nu_1} Z_2^{\nu_2} = 0,$$

for some $a_\nu \in k$. Since $E \cap V(Z_i) = p_i$, $i = 1, 2$, all a_ν but a_{300}, a_{111}, a_{021} and a_{012} have to vanish. This is obvious, because otherwise there would be multiple solutions, but the p_i are inflection points and so have multiplicity three. Hence E is given by

$$F = Z_1 Z_2 (\alpha Z_0 + \beta Z_1 + \gamma Z_2) + \delta Z_0^3 = 0,$$

for some $\alpha, \beta, \gamma, \delta \in k$. Now we have $\delta \neq 0$, because otherwise E would be reducible, a contradiction to the smoothness. Furthermore, it follows from $\mathbb{T}_{p_i}(E) = V(Z_i)$, $i = 1, 2$, that $\beta, \gamma \neq 0$, because

$$V(\gamma Z_1) = V(\sum_{i=0}^{2} \partial_i F(p_1) Z_i) = \mathbb{T}_{p_1}(E) = V(Z_1)$$

resp.

$$V(\beta Z_2) = V(\sum_{i=0}^{2} \partial_i F(p_2) Z_i) = \mathbb{T}_{p_2}(E) = V(Z_2).$$

Using the coordinate change $[Z_0 : Z_1 : Z_2] \longmapsto [\delta^{1/3} Z_0 : \beta Z_1 : \gamma Z_2]$, we may assume that the equation is of the following form:

$$F = Z_1 Z_2 (\alpha Z_0 + Z_1 + Z_2) + Z_0^3.$$

By calculating the common zeros of the derivatives of F, we remark that E is smooth if and only if $\alpha^3 + 27 \neq 0$. After passing to new coordinates $[Y_0 : Y_1 : Y_2]$ defined by

$$Z_0 = Y_0,$$

$$3Z_1 = -\alpha Y_0 - Y_1 + 2Y_2,$$
$$3Z_2 = -\alpha Y_0 + 2Y_1 - Y_2,$$

we can assume that E is given by

$$F = (27 + \alpha^3)Y_0^3 - 2Y_1^3 + 3Y_1^2Y_2 + 3Y_2^2Y_1 - 2Y_2^3 - 3\alpha Y_0(Y_1^2 - Y_1Y_2 + Y_2^2) = 0.$$

Let us denote by ϵ a primitive third root of unity. Note that this implies $\epsilon^2 + \epsilon + 1 = 0$. Then we remark that $(Y_1^2 - Y_1Y_2 + Y_2^2) = (Y_1 + \epsilon Y_2)(Y_1 + \epsilon^2 Y_2)$ and $-2Y_1^3 + 3Y_1^2Y_2 + 3Y_2^2Y_1 - 2Y_2^3 = -(Y_1 + \epsilon Y_2)^3 - (Y_1 + \epsilon^2 Y_2)^3$ holds and hence the cubic is given by

$$(27 + \alpha^3)Y_0^3 - (Y_1 + \epsilon Y_2)^3 - (Y_1 + \epsilon^2 Y_2)^3 - 3\alpha Y_0(Y_1 + \epsilon Y_2)(Y_1 + \epsilon^2 Y_2) = 0.$$

After a further change of coordinates and observing $\alpha^3 + 27 \neq 0$, we obtain that E is given by

$$(27 + \alpha^3)Z_0^3 - Z_1^3 - Z_2^3 - 3\alpha Z_0Z_1Z_2 = 0$$

and finally by

$$Z_0^3 + Z_1^3 + Z_2^3 + \lambda Z_0Z_1Z_2 = 0,$$

for some $\lambda \in k$. $\qquad\square$

Remark 1.1.6. *The normal form E_λ is not uniquely determined by E. For example, if one multiplies the coordinates by primitive third roots of unity, then λ is also multiplied by a primitive third root of unity, e.g. E_λ and $E_{\epsilon\lambda}$ can be carried into each other by collineations.*

The previous theorem also shows that the E_λ can be considered as a non-unique normal form of a smooth cubic curve. It is called the *Hesse normal form*. Although E_∞ is not smooth, we say that it is in Hesse normal form. The other singular members of the pencil can be found by means of the following

Lemma 1.1.7. *For $\lambda \in k$ let E_λ be a member of the Hesse pencil. Then it is smooth if and only if $\lambda^3 + 27 \neq 0$.*

Proof. By computing the common zeros of the partial derivatives, the assertion holds. $\qquad\square$

We also say that these additional three curves are in Hesse normal form. Therefore the space of cubic curves in Hesse normal form coincides with $\mathcal{P}$ and we will identify the two sets. The next corollary gives a very important geometrical property of the inflection points of any smooth cubic curve.

Corollary 1.1.8. *A line through two inflection points of a smooth cubic curve intersects the cubic in a third inflection point.*

Proof. We use the same notation as in Theorem 1.1.5. By the first part of its proof, we can assume that E is given by

$$Z_1 Z_2(\alpha Z_0 + \beta Z_1 + \gamma Z_2) + \delta Z_0^3 = 0.$$

The line $\overline{p_1, p_2} : Z_0 = 0$ intersects the cubic in the third point $s := [0 : 1 : -\beta/\gamma]$. Now we see, that $\alpha Z_0 + \beta Z_1 + \gamma Z_2 = 0$ is a inflection tangent line of s, because one can easily see that it intersects the cubic only in s. □

This corollary is the key point for the definition of the *Hesse configuration* in Chapter 1.2.

Let us now come back to the Hesse pencil. Therefore we fix any $E_\lambda \in \mathcal{P}$. By an easy computation, one finds that $He(E_\lambda)$ is given by

$$Z_0^3 + Z_1^3 + Z_2^3 + -\frac{108 + \lambda^3}{3\lambda^2} Z_0 Z_1 Z_2, \tag{1}$$

for $\lambda \neq 0, \infty$. Hence $He(E_\lambda)$ lies again in $\mathcal{P}$ and we can naturally define a map

$$He : \mathcal{P} \longrightarrow \mathcal{P}, \; E_\lambda \longmapsto He(E_\lambda).$$

We are now interested in the properties of this map and this leads us to

Lemma 1.1.9. *Let $\mathcal{P}$ denote the space of cubic curves in Hesse normal form. Then*

$$He : \mathcal{P} \longrightarrow \mathcal{P}, \; E_\lambda \longmapsto He(E_\lambda)$$

is well-defined and

i) $He(E_\lambda) = E_{-\frac{108+\lambda^3}{3\lambda^2}}$, $He(E_0) = E_\infty$ *and* $He(E_\infty) = E_\infty$. *In particular,* $He(E_\lambda) = E_\lambda$ *if and only if* $\lambda \in \{\infty, -3, -3\epsilon, -3\epsilon^2\}$, *where ϵ denotes a primitive third root of unity;*

ii) $He(E_\lambda)$ *is smooth if and only if* $\left(-\frac{108+\lambda^3}{3\lambda^2}\right)^3 + 27 \neq 0$.

iii) *The induced map on the space of parameters of the Hesse pencil*

$$\mathfrak{h} : \mathbb{P}^1 \longrightarrow \mathbb{P}^1,$$

$$\lambda \longmapsto -\frac{108 + \lambda^3}{3\lambda^2}, \; 0 \longmapsto \infty, \; \infty \longmapsto \infty$$

has degree three. Furthermore, its set of branching points coincide with $\{\infty, -3, -3\epsilon, -3\epsilon^2\}$, the fixed points of He. The corresponding ramification points are given by $\{0, 6, 6\epsilon, 6\epsilon^2\}$.

Proof. To prove *i*) we see by (1) that

$$He(E_\lambda) = E_{-\frac{108+\lambda^3}{3\lambda^2}}$$

and therefore $He(E_\lambda) \in \mathcal{P}$. In particular, the map is well-defined for $\lambda \neq 0, \infty$. Now we remark that

$$\lambda = -\frac{108+\lambda^3}{3\lambda^2} \iff \lambda^3 + 27 = 0 \iff \lambda \in \{-3, -3\epsilon, -3\epsilon^2\}.$$

By another computation, we get that $He(E_0) = E_\infty$ and $He(E_\infty) = E_\infty$ and so the equivalence holds and the map He is well-defined for all elements in $\mathcal{P}$.

To show *iii*) we compute the differential of $\mathfrak{h}$ and one finds that the ramification points correspond to the zeros of $\lambda(\lambda^3 - 6^3)$, which are given by $0, 6, 6\epsilon$ and $6\epsilon^2$. The corresponding branching points coincide with $\infty, -3, -3\epsilon$ and $-3\epsilon^2$. Note that each branching point has two points in its fibre. Since the map is given by a cubic equation, the fibres have a generic cardinality of three and therefore $\mathfrak{h}$ has degree three.

Assertion *ii*) is an immediate consequence of *i*) and Lemma 1.1.7. □

Let us denote by $\mathcal{P}_{reg}$ the smooth members of the Hesse pencil. The next Lemma will show that all elements in $\mathcal{P}_{reg}$ have the same inflection points.

Lemma 1.1.10. *For any $E_\lambda \in \mathcal{P}_{reg}$ holds*

$$He(E_\lambda) \cap E_\lambda = E_0 \cap E_\infty.$$

In particular, all smooth members of the pencil have the same inflection points. They are given by

$$p_{0,0} = [0:1:-1], \; p_{0,1} = [0:1:-\epsilon^2], \; p_{0,2} = [0:1:-\epsilon],$$
$$p_{1,0} = [1:0:-1], \; p_{1,1} = [1:0:-\epsilon], \; p_{1,2} = [1:0:-\epsilon^2],$$
$$p_{2,0} = [1:-1:0], \; p_{2,1} = [1:-\epsilon^2:0], \; p_{2,2} = [1:-\epsilon:0],$$

where ϵ denotes a primitive third root of unity. (Note that the intersection $E_0 \cap E_\infty$ is the set of base points of the Hesse pencil.)

Proof. Since $He(E_\lambda)$ and E_λ are in the Hesse pencil, by Lemma 1.1.9, it is obvious that $E_0 \cap E_\infty \subseteq He(E_\lambda) \cap E_\lambda$. Since E_λ is smooth, we have by Lemma 1.1.9 that $He(E_\lambda) \neq E_\lambda$. Hence we know by *Bézout's Theorem* (cf. [Ful08], Chap. 5.3] that $He(E_\lambda) \cap E_\lambda$ contains at most nine points. Furthermore, we easily verify that all $p_{i,j}, \; i, j = 0, 1, 2$, lie in $E_0 \cap E_\infty$ and thus the equality holds. By Lemma 1.1.3, the intersection $E_0 \cap E_\infty$ coincide with the set of all inflection points of all members of the pencil. □

Furthermore, we can by Lemma 1.1.7 compute the singular members of the Hesse pencil and we see that each one is the union of three lines:

$$E_\infty: \qquad Z_0 Z_1 Z_2 = 0,$$
$$E_{-3}: \qquad (Z_0 + Z_1 + Z_2)(Z_0 + \epsilon Z_1 + \epsilon^2 Z_2)(Z_0 + \epsilon^2 Z_1 + \epsilon Z_2) = 0,$$
$$E_{-3\epsilon}: \qquad (Z_0 + \epsilon Z_1 + Z_2)(Z_0 + \epsilon^2 Z_1 + \epsilon^2 Z_2)(Z_0 + Z_1 + \epsilon Z_2) = 0,$$
$$E_{-3\epsilon^2}: \qquad (Z_0 + \epsilon^2 Z_1 + Z_2)(Z_0 + \epsilon Z_1 + \epsilon Z_2)(Z_0 + Z_1 + \epsilon^2 Z_2) = 0.$$

We will call these four members of the pencil the *triangles* and denote them by $T_1, ..., T_4$, respectively. The singular points of them will be called the *vertices*. They are given by

$$v_0 = [1:0:0], \quad v_1 = [0:1:0], \quad v_2 = [0:0:1],$$
$$v_3 = [1:1:1], \quad v_4 = [1:\epsilon:\epsilon^2], \quad v_5 = [1:\epsilon^2:\epsilon],$$
$$v_6 = [1:\epsilon:1], \quad v_7 = [1:\epsilon^2:\epsilon^2], v_8 = [1:1:\epsilon],$$
$$v_9 = [1:\epsilon^2:1], v_{10} = [1:\epsilon:\epsilon], \quad v_{11} = [1:1:\epsilon^2].$$

The vertices are just mentioned here to get a complete picture of the setting. They will play an important role in Chapter 2.3, where we will define a dual Hesse pencil. Since the $p_{i,j}$ are nonsingular points, we remark that every $p_{i,j}$ is contained in every triangle, i.e. in four of the twelve lines. Moreover, every line of the triangles contains three of the $p_{i,j}$.

The Hesse pencil also induces a rational map

$$\mathbb{P}^2 \dashrightarrow \mathbb{P}^1, \ [Z_0 : Z_1 : Z_2] \longmapsto [Z_0 Z_1 Z_2 : Z_0^3 + Z_1^3 + Z_2^3],$$

which is not defined at the base points of the pencil. Let

$$\pi : S(3) \longrightarrow \mathbb{P}^2$$

be the blowing up of the base points. This defines a rational surface such that the composition of rational maps

$$S(3) \longrightarrow \mathbb{P}^2 \dashrightarrow \mathbb{P}^1$$

is a regular map

$$\Phi : S(3) \longrightarrow \mathbb{P}^1,$$

(cf. [Bea96, Theorem II.7]). It is easy to see that the fibre of Φ over a point $[\mu_1 : \mu_2] \in \mathbb{P}^1$ is isomorphic to the member of the Hesse pencil corresponding to $[t_0 : t_1] = [-\mu_2 : \mu_1]$ (cf. [Dol12, p. 134]). Furthermore, the map Φ defines a structure of a minimal elliptic surface on $S(3)$, which is a special case of an *elliptic modular surface $S(n)$* of level n (cf. [BH85, Chap. III]). The proof of this relies on methods far beyond the scope of this thesis.

1.2 The Hesse configuration

In this part of the thesis, we want to discuss the relation between the twelve lines of the decomposing singular members of the Hesse pencil and its nine base points, mentioned in the very last part of Chapter 1.1. To do this, we need the following

Definition 1.2.1. *A configuration $\mathcal{K}$ is a pair consisting of a set $\mathcal{M}$ and a set $\mathcal{L}$ of subsets of $\mathcal{M}$ such that two different elements of $\mathcal{L}$ intersect in at most one element. The elements of $\mathcal{M}$ are called* points *and the elements of $\mathcal{L}$ are called* lines. *Moreover we define the set $\mathcal{F}$ of* flags *of $\mathcal{K}$ to be all pairs (P, L) formed by a point P and a line L of $\mathcal{K}$ containing P.*

An isomorphism from $\mathcal{K}$ to $\mathcal{K}'$ is a bijection $\phi : \mathcal{M} \to \mathcal{M}'$ that induces a bijection between the set of flags.

A configuration $(\mathcal{M}, \mathcal{L})$ is called regular, *if the automorphism group acts transitively on the set of flags, and it is called* geometric *if there is a (projective or affine) plane E over some field k and a set of points $\mathcal{M}'$ and a set of lines $\mathcal{L}'$ in E such that $(\mathcal{M}, \mathcal{L})$ and $(\mathcal{M}', \mathcal{L}')$ are isomorphic.*

A configuration $\mathcal{K}$ is of type (p_γ, ℓ_π) if we have $p = |\mathcal{M}|$ and $\ell = |\mathcal{L}|$ and if each point lies on γ lines of $\mathcal{L}$ and each line contains π points of $\mathcal{M}$.

If we denote by $\pi_\mathcal{M}$ and $\pi_\mathcal{L}$ the canonical projections from $\mathcal{F}$ to $\mathcal{M}$ resp. to $\mathcal{L}$, we see that

$$\mathcal{F} = \bigsqcup_{P \in \mathcal{M}} \pi_\mathcal{M}^{-1}(P)$$

and

$$\mathcal{F} = \bigsqcup_{L \in \mathcal{L}} \pi_\mathcal{L}^{-1}(L).$$

this shows that a configuration of type (p_γ, ℓ_π) has exactly $p\gamma = \ell\pi$ flags.

Now we are able to give the following

Definition 1.2.2. *The* Hesse configuration *is the configuration $(9_4, 12_3)$ of the twelve lines of the decomposing singular members of the Hesse pencil and its nine base points.*

The next lemma will help us to get an idea of the geometrical properties of the Hesse configuration.

Lemma 1.2.3. *Let E be the affine plane over $\mathbb{F}_3$. Let $\mathcal{K}$ be the configuration given by all points and lines. Then we have :*

i) *This configuration is of type $(9_4, 12_3)$;*

ii) *The automorphism group is the affine group $AGL(2, \mathbb{F}_3)$ and the configuration is regular;*

iii) *Every configuration of type $(9_4, 12_3)$ is isomorphic to $\mathcal{K}$.*

12

Proof. Obviously, $AGL(2, \mathbb{F}_3)$ is contained in the automorphism group and it acts transitively on the flags. Therefore it suffices to consider the origin. There are four lines through the origin. Each line contains three points, hence the type is $(9_4, 12_3)$ and $i)$ is proved.

To prove $ii)$ let g be an automorphism of the configuration. Let L and L' be two different lines through the origin. Then gL and gL' are lines intersecting in a point P. Let $h_1 \in AGL(2, \mathbb{F}_3)$ be the translation from P to the origin such that $h_1 g$ fixes the origin. Then there exists $h_2 \in AGL(2, \mathbb{F}_3)$ such that L and L' are preserved under $h_2 h_1 g$. After multiplication with an appropriate $h_3 \in AGL(2, \mathbb{F}_3)$, we even get that $h_3 h_2 h_1 g$ acts trivial on L and L'. Then it also fixes all points of the four lines through the non-zero points $Q \in L$ and $R \in L'$ and thus $h_3 h_2 h_1 g$ has to be the identity and $g = h_1^{-1} h_2^{-1} h_3^{-1} \in AGL(2, \mathbb{F}_3)$.

To prove the uniqueness of the configuration choose a point and call it 1. Then the points in the lines through 1 can be numbered as follows:

$$L_1 := \{1, 2, 3\}, \ L_2 := \{1, 4, 5\}, \ L_3 := \{1, 6, 7\}, \ L_4 := \{1, 8, 9\}.$$

Since the first point was chosen arbitrary, we see that there is a line $\overline{I, J}$ through any different points I and J. The line $\overline{2, 4}$ contains $6, 7, 8$ or 9 as the third point. Up to renumbering we can assume that $L_5 = \{2, 4, 8\}$, hence the lines $\overline{4, 6}$ and $\overline{4, 7}$ can only contain 3 or 9 as the third point. Up to renumbering we get $L_6 = \{4, 6, 9\}$ and $L_7 = \{3, 4, 7\}$. The remaining lines are now uniquely determined as $L_8 = \{5, 7, 8\}$, $L_9 = \{3, 5, 9\}$, $L_{10} = \{2, 5, 6\}$, $L_{11} = \{3, 6, 8\}$ and $L_{12} = \{2, 7, 9\}$. This proves $iii)$. $\qquad\square$

Due to the last lemma, we can think of the *Hesse configuration* as follows:

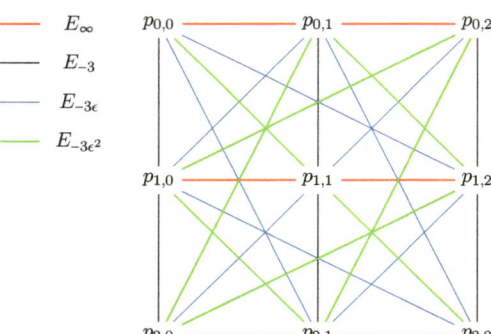

Figure 1: Hesse configuration

13

Remark 1.2.4. *The isomorphism between the configuration of the inflection points and the lines of the triangles and the points and lines in $\mathbb{F}_3^2$ is given by*

$$p_{i,j} \longmapsto (i,j), \; i,j = 0,1,2.$$

1.3 The Hesse group

In this part of the thesis, we want to study the subgroup of $PGL(3,k)$ leaving the Hesse pencil invariant. It is called the *Hesse group* and is denoted by G_{216}, due to its order.

For the sake of simplicity we will use the following notation:

$$\mathcal{P} = \{E_\lambda : \lambda \in k\} \cup \{E_\infty\}, \mathcal{I} := \{p_{i,j} : i,j = 0,1,2\},$$

$$\mathcal{I}(E_\lambda) := \{\text{Inflection points of a smooth } E_\lambda\}.$$

Then G_{216} can be described as the subgroup of $PGL(3,k)$ fulfilling one of the following conditions:

Lemma 1.3.1. *For $g \in PGL(3,k)$ is equivalent :*

i) $g(\mathcal{P}) = \mathcal{P}$;

ii) $g(E_\lambda) \in \mathcal{P}$ for any $E_\lambda \in \mathcal{P}$;

iii) $g(\mathcal{I}) = \mathcal{I}$.

Proof. $i) \Rightarrow ii)$ This is obvious.

$ii) \Rightarrow iii)$ Since $g(E_\lambda) = E_\mu$ for any $\mu \in k \cup \{\infty\}$, we obtain

$$\mathcal{I} = \mathcal{I}(E_\mu) = \mathcal{I}(g(E_\lambda)) = g(\mathcal{I}(E_\lambda)) = g(\mathcal{I}).$$

$iii) \Rightarrow i)$ Let $\mathcal{P}_{reg}$ denote as used before the set of smooth members of $\mathcal{P}$. Let $E_\lambda \in \mathcal{P}_{reg}$ and therefore $g(E_\lambda) \in \mathcal{P}_{reg}$. By assumption we have

$$\mathcal{I}(g(E_\lambda)) = g(\mathcal{I}(E_\lambda)) = g(\mathcal{I}) = \mathcal{I}$$

and thus $g(E_\lambda) \in \mathcal{P}_{reg}$ holds. Since $\mathcal{P}_{reg} \subseteq \mathcal{P}$ is dense, we obtain

$$g(\mathcal{P}) = g(\overline{\mathcal{P}_{reg}}) \subseteq \overline{g(\mathcal{P}_{reg})} = \overline{\mathcal{P}_{reg}} = \mathcal{P}.$$

If we argue analogously for g^{-1}, we get the other inclusion. $\qquad\square$

We want to describe G_{216} in detail. Since every element in the group leaves the Hesse configuration invariant, we obtain by Lemma 1.2.3 that G_{216} lies in $AGL(2, \mathbb{F}_3)$. Moreover, if an element of G_{216} is trivial, considered as element in $AGL(2, \mathbb{F}_3)$, we know that it also fixes four lines in general position, e.g. the lines $Z_0 = 0, Z_1 = 0, Z_2 = 0$ and $Z_0 + Z_1 + Z_2 = 0$. Thus it has to be the identity and we have shown, that $G_{216} \subseteq AGL(2, \mathbb{F}_3)$ as subgroup.

Since every $g \in G_{216}$ permutes the members of the Hesse pencil, we also can think of any $g \in G_{216}$ as an element of $Aut(\mathbb{P}^1)$ via the homomorphism

$$\Phi : G_{216} \longrightarrow Aut(\mathbb{P}^1).$$

Indeed we obtain the following

Theorem 1.3.2. *The* Hesse *group is isomorphic to the special affine linear group of* $\mathbb{F}_3^2$.

To prove this, we define the following collineations in $PGL(3, k)$:

$$A := \begin{pmatrix} 0 & 0 & 1 \\ 1 & 0 & 0 \\ 0 & 1 & 0 \end{pmatrix}, \ B := \begin{pmatrix} 1 & 0 & 0 \\ 0 & 0 & 1 \\ 0 & 1 & 0 \end{pmatrix}, \ C := \begin{pmatrix} 1 & 0 & 0 \\ 0 & \epsilon & 0 \\ 0 & 0 & \epsilon^2 \end{pmatrix},$$

$$D := \begin{pmatrix} 1 & 0 & 0 \\ 0 & \epsilon & 0 \\ 0 & 0 & \epsilon \end{pmatrix}, \ E := \begin{pmatrix} 1 & 1 & 1 \\ 1 & \epsilon & \epsilon^2 \\ 1 & \epsilon^2 & \epsilon \end{pmatrix}$$

(cf. [BK86, p. 297], where ϵ denotes as usual a primitive third root of unity. It is obvious that A, B and C lie in G_{216}. D maps any member E_λ of the pencil to $E_{\epsilon\lambda}$ and thus lies in G_{216}. The collineation E transforms E_λ to $E_{\lambda+3,18-3\lambda}$ for $\lambda \in k$ and E_∞ to E_{-3} and thus also lies in G_{216}. Furthermore it is obvious that $A, B, C \in \ker(\Phi)$.

To study G_{216}, we have to understand the action on the four singular members of the Hesse pencil. Therefore we prove the following

Lemma 1.3.3. *If $g \in PGL(3, k)$ satisfies $g(E_\infty) = E_\infty$ and $g(E_{-3}) = E_{-3}$, then we obtain that $g \in \langle A, B, C \rangle$ and therefore $g \in \ker(\Phi)$. In particular $g(E_{-3\epsilon}) = E_{-3\epsilon}$ and $g(E_{-3\epsilon^2}) = E_{-3\epsilon^2}$.*

Proof. Since g leaves E_∞ invariant, it has to be a product of a homothety and a permutation. Due to the fact that $\langle A, B \rangle \cong \mathfrak{S}_3$, we can assume that g is a homothety, hence

$$g = \begin{pmatrix} 1 & 0 & 0 \\ 0 & a & 0 \\ 0 & 0 & b \end{pmatrix},$$

for $a, b \in k$. Since g also leaves

$$E_{-3} : (Z_0 + Z_1 + Z_2)(Z_0 + \epsilon Z_1 + \epsilon^2 Z_2)(Z_0 + \epsilon^2 Z_1 + \epsilon Z_2) = 0$$

invariant and thus permutes the lines

$$\ell_1 := (Z_0 + Z_1 + Z_2) = 0,$$

$$\ell_2 := (Z_0 + \epsilon Z_1 + \epsilon^2 Z_2) = 0,$$

$$\ell_3 := (Z_0 + \epsilon^2 Z_1 + \epsilon Z_2) = 0,$$

we get the following cases:

$$g(\ell_1) = \ell_1 \Rightarrow g = id \in PGL(3, k);$$
$$g(\ell_1) = \ell_2 \Rightarrow g = C \in PGL(3, k);$$
$$g(\ell_1) = \ell_3 \Rightarrow g = C^2 \in PGL(3, k).$$

This completes the proof. □

In particular, we obtain that the kernel of Φ is equal to $\langle A, B, C \rangle$, because every element in the kernel of Φ leaves E_∞ and E_{-3} invariant and thus lies in $\langle A, B, C \rangle$ by the lemma. Since the other inclusion is obvious, we get

$$\ker(\Phi) = \langle A, B, C \rangle = \langle A, C \rangle \rtimes \langle B \rangle,$$

since $\langle A, C \rangle$ is a normal subgroup. In particular, we also see that $|\ker \Phi| = 18$.

Now every element of G_{216} acts on the pencil by permuting its four singular members and thus we get an homomorphism

$$\Psi : G_{216} \longrightarrow \mathfrak{S}_4.$$

By Lemma 1.3.3 we get $\ker(\Phi) = \ker(\Psi)$ and therefore $\mathrm{im}(\Phi) \cong \mathrm{im}(\Psi)$. Indeed we have

$$\mathrm{im}(\Psi) = \mathfrak{A}_4. \tag{2}$$

Namely, $\Psi(D) = (1\,2\,3)$ and $\Psi(E) = (1\,2)(3\,4)$ generate $\mathfrak{A}_4$. For $\mathrm{im}(\Psi) = \mathfrak{S}_4$ there would be a $g \in G_{216}$ fixing E_∞ and E_{-3} but permuting $E_{-3\epsilon}$ and $E_{-3\epsilon^2}$. This is impossible by Lemma 1.3.3. So we get $|G_{216}| = 18 \cdot 12 = 216$ and G_{216} is generated by A, B, C, D and E. We conclude that G_{216} is a subgroup of index two in $AGL(2, \mathbb{F}_3)$. Therefore every element of order three lies in G_{216}, hence $\mathbb{F}_3^2 \subseteq G_{216}$. Note that $\mathbb{F}_3^2$ is generated by A and C. Now it suffices to show that $G_{216}/\mathbb{F}_3^2 = SL(2, \mathbb{F}_3) \subseteq GL(2, \mathbb{F}_3)$. Since $G_{216}/\mathbb{F}_3^2$ is again of index two in $GL(2, \mathbb{F}_3)$, we conclude that $G_{216}/\mathbb{F}_3^2 = SL(2, \mathbb{F}_3)$, because it is well-known that $SL(2, \mathbb{F}_3)$ is the only subgroup of index two in $GL(2, \mathbb{F}_3)$. Hence $G_{216} \cong ASL(2, \mathbb{F}_3) \cong \mathbb{F}_3^2 \rtimes SL(2, \mathbb{F}_3)$. This shows Theorem 1.3.2.

1.4 An algebraic approach

Let us now construct the Hesse configuration without using the Hesse pencil. Let E be an elliptic curve. We want to define a group law on E. Therefore we fix one of the inflection points p_0. Then we define $(E, +)$ by means of the following rule:

Definition 1.4.1. Let $p, q \in E$, ℓ the line connecting p and q, respectively the tangent line to E if $p = q$. Then $p + q = r$ if and only if r, p_0 and the third point of intersection in $\ell \cap E$ are collinear.

16

It follows by Proposition 2.2 in [Sil86, Chap. III, §2] that this defines a commutative group law on E with origin p_0. Note that we have $p + q + r = 0$ if and only if p, q and r are collinear. Let us now consider the set G of inflection points on E. We want to show that this forms a subgroup of E, which coincide with the group of 3-torsion points. Moreover, we want to show that G is abstractly isomorphic to $(\mathbb{Z}/3\mathbb{Z})^2$. The group of 3-torsion points is denoted by $E[3]$. By Definition 1.4.1, each inflection point is a 3-torsion point and thus $G \subseteq E[3]$ as set. The set of inflection points G has nine elements (cf. Lemma 1.1.10), each one of order three. First, we show that G is a subgroup of $(E, +)$. Note that we have already done this in Corollary 1.1.8, but the next proof works just by means of Definition 1.4.1.

Lemma 1.4.2. *Any line $\overline{p, q}$ through two inflection points intersects E at another inflection point r, i.e. G is a subgroup of $E[3]$. Furthermore p, q and r form a coset with respect to some subgroup of G.*

Proof. We have $p + p = -p$ and $q + q = -q$, because p and q are inflection points. Let $\overline{p, q}$ intersect E at $-p - q =: r$. Then $r + r = -2p - 2q = p + q = -r$ and therefore r is an inflection point, since it has multiplicity three. Let us define $H := \{p, q, -p - q\}$. If $p = 0$, H is already a subgroup. Otherwise $\tilde{H} := \{0, q - p, p - q\}$ is a subgroup and $H = p + \tilde{H}$. $\square$

Note that $-r$ also is an inflection point, since we have $-r - r - r = -3r = 0$. Therefore G is a subgroup of $(E, +)$ and since every element is of order three, it is even a subgroup of $E[3]$. Clearly, every 3-torsion point is an inflection point and thus lies in G. Hence we have shown that $G = E[3]$. Since G is a group with nine elements of order three, we conclude $G = E[3] \cong (\mathbb{Z}/3\mathbb{Z})^2$. Therefore G contains four subgroups of order three such that we have at most twelve lines, passing through three inflection points. On the other hand, if $p + H = \tilde{p} + \tilde{H}$ we get $H = \tilde{H}$ and so have get exactly twelve lines, each containing three inflection points. They are called the *inflection lines*, not to mix them up with the inflection tangent lines. Since each element in G is contained in four subgroups, we conclude that each inflection point is contained in four inflection lines. This gives again the definition of the Hesse configuration $(9_4, 12_3)$ (cf. Definition 1.2.2), but without having considered the Hesse pencil. This configuration is of course independent of the choice of p_0, since it was chosen arbitrary.

If we think of the triangles in the end of Chapter 1.1, we get the following

Remark 1.4.3. *If we fix a point $p_{i,j}$ as the origin of the group law, then the side of a triangle T_l passing through $p_{i,j}$ contains three base points forming a subgroup of order three, while the other sides of T_l contain the cosets with respect to this subgroup.*

1.5 The Weierstrass normal form and the j-invariant

In this part we assume, additionally to Chapter 1.1, that the characteristic of k is different from two. Since we are also interested in other properties of the Hesse pencil, e.g. the j-invariant, we would like to compute a *Weierstrass normal form*. This leads us to the next theorem. For aesthetic reasons, we consider now $E_{-3\lambda}$ instead of E_λ.

Theorem 1.5.1. *Let $E_{-3\lambda} \in \mathcal{P}$. Then it is isomorphic to the curve given by*

$$x_2^2 = 4x_1^3 - 24(1 + (\frac{1}{2}\lambda)^3)x_1 - 8(1 + 10(\frac{\lambda^3}{2} - 8\frac{\lambda^6}{2^6})).$$

In particular, its j-invariant is given by

$$j(E_{-3\lambda}) = 27\Big(\frac{\lambda(\lambda^3 + 8)}{\lambda^3 - 1}\Big)^3,$$

(cf. [PP09, Proposition 2.2]).

Proof. We expand the proof of Proposition 2.3 in [PP09]. Let us fix the inflection point $[1:-1:0]$ with inflection line

$$Z_0 + Z_1 + \lambda Z_2 = 0.$$

We introduce new coordinates given by

$$X_0 = Z_0 + Z_1 + \lambda Z_2$$

$$X_1 = \frac{1}{2}(Z_0 + Z_1)$$

$$X_2 = \frac{1}{2}(Z_0 - Z_1).$$

Since this is equivalent to

$$Z_0 = X_1 + X_2$$

$$Z_1 = X_1 - X_2$$

$$Z_2 = \frac{1}{\lambda}(X_0 - 2X_1),$$

$E_{-3\lambda}$ is transformed to

$$8X_1^3 + \frac{1}{\lambda^3}(X_0 - 2X_1)^3 - 3(X_1^2 - X_2^2)X_0 = 0.$$

By using the affine coordinates $x_1 = \frac{X_1}{X_0}$ and $x_2 = \frac{X_2}{X_0}$, we get to the affine equation

$$x_2^2 = \frac{8}{3}(\frac{1}{\lambda^3} - 1)x_1^3 + (1 - \frac{4}{\lambda^3})x_1^2 + \frac{2}{\lambda^3}x_1 - \frac{1}{3\lambda^3}.$$

By another change of coordinates and a *Tschirnhaus-Reduction*, i.e.

$$x_1 \mapsto x_1 - \frac{a}{3},$$

where a denotes the coefficient of x_1^2, one gets the claimed form. The formula for the j-invariant follows by Proposition 2.2 in [PP09]. $\qquad\square$

We want to give an interesting application of the j-invariant to the Hesse pencil. By Theorem 10 [BK86, Chap. 7.3], $E_{-3\lambda}$ and $E_{-3\mu}$ are isomorphic via a collineation if and only if $j(E_{-3\lambda}) = j(E_{-3\mu})$. By Chapter 1.3, we know that these collineations are given by the Hesse group G_{216}. To be more precisely, we even know by Chapter 1.3 (2) that it suffices to consider the tetrahedral group $\mathfrak{A}_4$, since the other elements of G_{216} preserve every member of the Hesse pencil. If we consider now the action of $\mathfrak{A}_4$ on the Hesse pencil, we conclude that the orbits correspond to the set of curves with the same j-invariant. In particular, the j-invariant induces a map on the space of parameters of the smooth members of the Hesse pencil, which has degree twelve:

$$j : \mathbb{P}^1 \to \mathbb{P}^1, \ -3\lambda \mapsto 27\Big(\frac{\lambda(\lambda^3 + 8)}{\lambda^3 - 1}\Big)^3.$$

2   The Cayleyan curve

In this section, we want to understand the construction of the *Cayleyan curve* of a smooth cubic curve E, which is a curve in $\check{\mathbb{P}}^2$. Futhermore, we will consider the dual Hesse pencil and associate to a smooth cubic curve E_λ the correspoding *Cayleyan curve* $Ca(E_\lambda)$. We will even see that this is a member of the dual Hesse pencil. To do this, we need the theory of *polar hypersurfaces*. Moreover, we assume in this chapter that the characteristic of k is zero.

2.1   Polar hypersurfaces

We follow now Chapter 1.1.1 and Chapter 1.1.2 of [Dol12] and fill out the details. In the following we denote by $X = V(F) \subset \mathbb{P}^n$ a hypersurface, where F is a homogeneous polynomial of degree d. Eventually, we will just need the case $n = 2$ and $d = 3$, but the proofs are almost the same and so we do the theory for the arbitrary case. Furthermore, we denote by $Sing(X)$ the singular points of X. The *Hessian* is defined to be $V(\det((\partial_i\partial_j F)_{i,j=1,\ldots n}))$, analogous to Definition 1.1.2. The tangent space of a point $y \in \mathbb{P}^n$ with respect to X is denoted by $\mathbb{T}_y(X)$, analogous to chapter one. It is given by $V(\sum_{i=0}^n \partial_i F(y)Z_i)$ for homogeneous coordinates $[Z_0 : \ldots : Z_n]$.
Let $a = [a_0 : \ldots : a_n] \in \mathbb{P}^n$. For $\mathbf{i} \in \mathbb{N}^{n+1}$ and $k \in \mathbb{N}$ we define the vector

notation:

$$\mathbf{i}! := i_0! \dots i_n!, \quad \binom{k}{\mathbf{i}} := \frac{k!}{\mathbf{i}!}, \quad |\mathbf{i}| := i_0 + \dots + i_n \ , \partial^{\mathbf{i}} := \partial_0^{i_0} \dots \partial_n^{i_n}, \quad a^{\mathbf{i}} := a_0^{i_0} \dots a_n^{i_n}.$$

This leads us to

Definition 2.1.1. *Let* $X = V(F)$ *be a hypersurface of degree* d, $\mathbf{i} \in \mathbb{N}^{n+1}$, $k \in \mathbb{N}$, $a = [a_0 : \dots : a_n] \in \mathbb{P}^n$. *Then the hypersurface*

$$P_{a^k}(X) := V(D_{a^k}(F))$$

of degree $d - k$ *is called the* k-*th polar hypersurface (or just the* k-*th polar) of the point* a *with respect to the hypersurface* X, *where*

$$D_{a^k}(F) = \sum_{|\mathbf{i}|=k} \binom{k}{\mathbf{i}} a^{\mathbf{i}} \partial^{\mathbf{i}} F.$$

This definition is of course independent of the choice of the representative of $a \in \mathbb{P}^n$, since it is defined by a zero locus. Clearly, the 0-th polar of X is X, independent of a. The next observation will give us a very useful symmetry of the polars.

Lemma 2.1.2. *For* $a, b \in \mathbb{P}^n$ *we have:*

$$D_{a^k}(D_{b^l}(F)) = D_{b^l}(D_{a^k}(F)),$$

Proof. By easy calculations we get:

$$
\begin{aligned}
D_{a^k}(D_{b^l}(F)) &= \sum_{|\mathbf{i}|=k} \binom{k}{\mathbf{i}} a^{\mathbf{i}} \partial^{\mathbf{i}} \sum_{|\mathbf{j}|=l} \binom{l}{\mathbf{j}} b^{\mathbf{j}} \partial^{\mathbf{j}} F \\
&= \sum_{|\mathbf{i}|=k, |\mathbf{j}|=l} \binom{k}{\mathbf{i}} \binom{l}{\mathbf{j}} a^{\mathbf{i}} b^{\mathbf{j}} \partial^{\mathbf{i}} \partial^{\mathbf{j}} F \\
&= \sum_{|\mathbf{j}|=l} \binom{l}{\mathbf{j}} b^{\mathbf{j}} \partial^{\mathbf{j}} \sum_{|\mathbf{i}|=k} \binom{k}{\mathbf{i}} a^{\mathbf{i}} \partial^{\mathbf{i}} F \\
&= D_{b^l}(D_{a^k}(F)).
\end{aligned}
$$

$\square$

In the following we will connect the polars with the Hessian. Let us start with some technical

Lemma 2.1.3. *For any* $0 \le k \le d$, *we have*

$$a \in P_{a^k}(X) \iff a \in X.$$

Proof. This is an immediate consequence of the *Euler formula* :

$$d \cdot F(Z_0, \ldots, Z_n) = \sum_{i=0}^{n} \partial_i F(Z_0, \ldots, Z_n) Z_i.$$

To prove this equality, we use that for all $t \in k$ holds

$$F(tZ_0, \ldots, tZ_n) = t^d F(Z_0, \ldots, Z_n).$$

Therefore we get :

$$\partial_t F(tZ_0, \ldots, tZ_n) = d \cdot t^{d-1} F(Z_0, \ldots, Z_n).$$

On the other hand we get by the chain rule :

$$\partial_t F(tZ_0, \ldots, tZ_n) = \sum_{i=0}^{n} \partial_i F(tZ_0, \ldots, tZ_n) Z_i.$$

The *Euler formula* follows by setting $t = 1$.

Applying this to the partial derivatives, we obtain

$$d \cdot (d-1) \cdot \ldots \cdot (d-k+1) \cdot F(Z_0, \ldots, Z_n) = \sum_{|\mathbf{i}|=k} \binom{k}{\mathbf{i}} Z^{\mathbf{i}} \partial^{\mathbf{i}} F(Z_0, \ldots, Z_n).$$

By setting $Z = a$, we have

$$D_{a^k}(F) = \sum_{|\mathbf{i}|=k} \binom{k}{\mathbf{i}} a^{\mathbf{i}} \partial^{\mathbf{i}} F.$$

We conclude

$$a \in X \iff F(a) = 0 \iff D_{a^k}(F)(a) = 0 \iff a \in P_{a^k}(X),$$

for any $0 \le k \le d$. $\qquad\square$

In particular, $d! F = D_{a^d}(F)$ and thus $P_{a^d}(X) = \mathbb{P}^n$ if $a \in X$, and $P_{a^d}(X) = \varnothing$ else. Using this, we get the symmetry of

Lemma 2.1.4. *For $a, b \in \mathbb{P}^n$, we have*

$$a \in P_{b^k}(X) \iff b \in P_{a^{d-k}}(X).$$

Proof. We use the last observation and Lemma 2.1.2. Let $a \in P_{b^k}(X)$. Then

$$\mathbb{P}^n = P_{a^{d-k}}(P_{b^k}(X)) = V(D_{a^{d-k}}(D_{b^k}(F)))$$
$$= V(D_{b^k}(D_{a^{d-k}}(F))) = P_{b^k}(P_{a^{d-k}}(X))$$

and therefore $b \in P_{a^{d-k}}(X)$. For $b \in P_{a^{d-k}}(X)$, we conclude $a \in P_{b^k}(X)$ analogously. $\qquad\square$

This lemma gives a very important relation between the $(d-k)$-th polar and the k-th polar. In this chapter we study this relation for $k = 1$. In Chapter 2.2 we will consider the case $k = 2$.

The next theorem will help us to get some idea of the geometry of the $(d-1)$-th polar and the first polar, which are related to the tangent space of X as follows:

Theorem 2.1.5. *Let $a \in X$. If $a \notin Sing(X)$, then we have $P_{a^{d-1}}(X) = \mathbb{T}_a(X)$. If $a \in Sing(X)$, we obtain $P_{a^{d-1}}(X) = \mathbb{P}^n$. Moreover, for any $x \in \mathbb{P}^n$,*

$$(X \cap P_x(X)) \smallsetminus Sing(X) = \{y \in X : x \in \mathbb{T}_y(X)\}.$$

Proof. By Lemma 2.1.4, we have $s \in P_{a^{d-1}}(X)$ if and only if $a \in P_s(X)$, i.e. $D_s(F)(a) = 0$. For $a \notin Sing(X)$, this holds if and only if $s \in \mathbb{T}_a(X)$, by the definition of the tangent space. For $a \in Sing(X)$, we have $D_s(F) = 0$ and the assertion is certainly true.

The second part is easy, because $y \in X \smallsetminus Sing(X)$ implies that $y \in P_x(X)$ if and only if $x \in \mathbb{T}_y(X)$, as shown above. $\qquad\square$

Let $P_{a^{d-1}}(X) = \mathbb{P}^n$. Then we have by Lemma 2.1.3 $a \in X$ and by Theorem 2.1.5 even $a \in Sing(X)$. Therefore we have shown the following

Lemma 2.1.6. *Let $a \in \mathbb{P}^n$. Then we have*

$$P_{a^{d-1}}(X) = \mathbb{P}^n \iff a \in Sing(X).$$

Applying this lemma to $P_{a^{d-2}}(X)$, we find

$$b \in Sing(P_{a^{d-2}}(X)) \iff P_b(P_{a^{d-2}}(X)) = \mathbb{P}^n \iff D_b(D_{a^{d-2}}(F)) = 0,$$

which proves the following

Lemma 2.1.7. *Let $a \in \mathbb{P}^n$. Then*

$$Sing(P_{a^{d-2}}(X)) = \{b \in \mathbb{P}^n : D_b(D_{a^{d-2}}(F)) = 0\}.$$

The $(d-2)$-th polar gives an important characterization of the Hessian and leads us to the next section.

2.2 Polar quadrics and the Hessian

For this section, we follow Chapter 1.1.3 and Chapter 1.1.4 of [Dol12]. Let us recall that the Hessian is given by $V(\det((\partial_i\partial_j F)_{i,j=1,...n}))$. Furthermore, we use the same notation as in Chapter 2.1. of this thesis. The *polar quadric* is defined as follows:

Definition 2.2.1. *The $(d-2)$-th polar of X with respect to $a \in \mathbb{P}^n$ is called the* polar quadric *of X with respect to a.*

Since the $(d-2)$-th polar is always of degree two, the name of the polar quadric is justified. By the symmetry of Lemma 2.1.4 we get

$$D_{a^{d-2}}(F)(x) = 0 \iff x \in P_{a^{d-2}}(X) \iff a \in P_{x^2}(X) \iff D_{x^2}(F)(a) = 0,$$

for $x \in \mathbb{P}^n$. Therefore, if the polar quadric is defined by the quadratic form $q(x) = D_{a^{d-2}}(F)(x)$, we obtain that

$$q(x) = D_{x^2}(F)(a) = \sum_{|\mathbf{i}|=2} \binom{2}{\mathbf{i}} x^{\mathbf{i}} \partial^{\mathbf{i}} F(a). \tag{3}$$

This observation gives the very important description of the Hessian by means of the next

Lemma 2.2.2. *The Hessian $He(X)$ is the locus of all points $a \in \mathbb{P}^n$ such that $P_{a^{d-2}}(X)$ is singular.*

Proof. Let $v \in Sing(P_{a^{d-2}}(X))$. This holds if and only if

$$\partial_j D_{a^{d-2}}(F)(v) = 0,$$

for all $0 \le j \le n$. Using the equation of (3), this is equivalent to

$$\partial_j \sum_{|\mathbf{i}|=2} \binom{2}{\mathbf{i}} v^{\mathbf{i}} \partial^{\mathbf{i}} F(a) = 0,$$

for all $0 \le j \le n$. This can be sorted as follows:

$$\sum_{i=0}^{n} 2v_i \partial_j \partial_i F(a) = 0,$$

for all $0 \le j \le n$. Finally, this is equivalent to $a \in He(X)$, because the last equation gives a non-trivial linear combination of the columns of the Hesse matrix, i.e. $\det((\partial_i \partial_j F(a))_{i,j=1,\ldots n}) = 0$. $\qquad\square$

We also want to give an additional equivalent description of the Hessian, based on the properties of the first polars.

Lemma 2.2.3. *The Hessian $He(X)$ is the locus of singular points of all first polars with respect to X, i.e.*

$$He(X) = \bigcup_{a \in \mathbb{P}^n} Sing(P_a(X)).$$

Proof. Let $a \in He(X)$. Then we have a non-trivial linear combination of the columns of the Hesse matrix, hence there exists some $b \in \mathbb{P}^n$ such that $a \in Sing(P_b(X))$. Conversely, let $a \in Sing(P_b(X))$ for some $b \in \mathbb{P}^n$. Then b induces a non-trivial linear combination of the colums of the Hesse matrix, i.e. $\det((\partial_i \partial_j F(a))_{i,j=1,\ldots n}) = 0$. We conclude that $a \in He(X)$. $\qquad\square$

For the special case of hypersurfaces of degree three the first polar coincides with the $(d-2)$-th polar and therefore one can use these both descriptions. This will be very useful in Chapter 2.3. Another important property of the polar quadric with respect to a is the following

Lemma 2.2.4. *The polar quadric of X with respect to a nonsingular point $a \in X$ is tangent to X at a.*

Proof. The proof is an immediate consequence of Theorem 2.1.5. We have

$$\mathbb{T}_a(P_{a^{d-2}}(X)) = P_a(P_{a^{d-2}}(X)) = P_{a^{d-1}}(X) = \mathbb{T}_a(X).$$

$\square$

We have learned enough about the polars and the connection to the Hessian and can proceed to the construction of the *Cayleyan curve*.

2.3   The dual Hesse pencil and the Cayleyan of a plane cubic

For this part of the thesis we have used the skeleton of section three in [AD10]. We fill out the details by ideas of section three in [Kuw11] and some ideas of ourselves. Now we consider a plane curve E given by $V(F)$, where $F(Z_0, Z_1, Z_2)$ is a homogeneous polynomial of degree three. The first polar with respect to $q = [a : b : c] \in \mathbb{P}^2$ is as usual denoted by $P_q(E)$ and defined to be $a\partial_0 F + b\partial_1 F + c\partial_2 F = 0$. Since it coincides with the polar quadric, we sometimes will say polar quadric instead of first polar. By Lemma 2.2.2, we see that the Hessian curve of E coincides with the locus of points q such that the polar quadric $P_q(E)$ is the union of two lines, which are a priori not distinct. Now fix some member E_λ of the Hesse pencil and let $p_{i,j} = [a : b : c] \in \mathbb{P}^2$ be one of the base points of the pencil, which are given in Lemma 1.1.10. Hence the first polar $P_{p_{i,j}}(E_\lambda)$ is reducible, i.e. it is the union of two lines. By Lemma 2.2.4, one of the two lines is the tangent line $\mathbb{T}_{p_{i,j}}(E_\lambda)$. After an easy computation, we find that the other line is given by

$$L_{i,j} : aZ_0 + bZ_1 + cZ_2 = 0.$$

The lines $L_{i,j}$, $i, j = 0, 1, 2$, are called the *harmonic polars*. By Theorem 2.1.5 we obtain, that the line $L_{i,j}$ intersects the curve E_λ at three points q_k such that $p_{i,j} \in \mathbb{T}_{q_k}(E_\lambda)$. Moreover, if we fix $p_{i,j}$ as the origin of the group law, we can prove the following

Lemma 2.3.1. *The group of 2-torsion points $E_\lambda[2]$ is equal to the group $G := \{p_{i,j}, q_1, q_2, q_3\}$.*

Proof. This can be done without general theory about the n-torsion group of an elliptic curve. First, we want to prove that G is a subgroup of $E_\lambda[2]$. Since we have $p_{i,j} \in \mathbb{T}_{q_k}(E_\lambda)$, every point q_k has order two. Furthermore, every line through two of the q_k coincides with $L_{i,j}$, and therefore G is closed under addition. This proves that G is a group and $G \subseteq E[2]$ as subgroup. Conversely, let $x \in E_\lambda[2]$, i.e. we have $p_{i,j} \in \mathbb{T}_x(E_\lambda)$. By Theorem 2.1.5 and Lemma 2.1.4, this implies $x \in P_{p_{i,j}}(E_\lambda)$. Thus we conclude $x \in \mathbb{T}_{p_{i,j}}(E_\lambda) \cup L_{i,j}$ if and only if $x = p_{i,j}$ or $x = q_k$, for $k = 1, 2, 3$. $\qquad\square$

Remark 2.3.2. *Projecting from the point $p_{i,j}$ we exhibit each curve of the Hesse pencil as a double cover of $\mathbb{P}^1$ branched at four points. The corresponding ramification points coincide with the group of 2-torsion points, namely by Lemma 2.3.1 with $p_{i,j}, q_1, q_2$ and q_3.*

In Lemma 1.1.9 we have defined the map

$$He : \mathcal{P} \longrightarrow \mathcal{P}, \ E_\lambda \longmapsto He(E_\lambda)$$

and we have shown that $He(E_\lambda) = E_{-\frac{108+\lambda^3}{3\lambda^2}}$. The next lemma will give a geometrical interpretation of $He^{-1}(E_\lambda)$ for any $\lambda \in k$.

Lemma 2.3.3. *Let E_λ be a nonsingular member of the Hesse pencil and $L_{i,j}$ any harmonic polar. Let $L_{i,j} \cap E_\lambda = \{q_1, q_2, q_3\}$ and E_{λ_k}, $k=1,2,3$, be the curve from the Hesse pencil whose tangent at the base point $p_{i,j}$ contains q_k. Then $He(E_\mu) = E_\lambda$ if and only if $\mu \in \{\lambda_1, \lambda_2, \lambda_3\}$.*

Proof. We refer to Proposition 3.2 of [AD10]. $\qquad\square$

Let us now consider the harmonic polars as points in the dual plane $\check{\mathbb{P}}^2$. They give the set of base points of a Hesse pencil in $\check{\mathbb{P}}^2$. Clearly, if we identify the plane with its dual by means of the quadratic form $Z_0^2 + Z_1^2 + Z_2^2$, the equation of the dual Hesse pencil coincides with the equation of the original pencil, since the coefficients of the harmonic polars coincide with the coordinates of the base points of the original pencil.

Now we define the dual Hesse configuration to be the configuration consisting of the nine points dual to the harmonic polars and the twelve lines dual to the vertices of the singular members of the original pencil. Considered as configuration in $\check{\mathbb{P}}^2$, we easily see that it is of type $(9_4, 12_3)$ and thus we can think of it as in Figure 1. If we consider the nine harmonic polars and the twelve vertices as configuration in $\mathbb{P}^2$, we note that each point is contained in three lines and each line consists of four points, hence the configuration is of type $(12_3, 9_4)$.

Finally, we want to associate to a smooth member E_λ a curve in the dual plane $\check{\mathbb{P}}^2$. Therefore let $E = V(F)$ be a smooth plane cubic curve such that $He(E)$ is smooth. If the Weierstrass normal form of E in affine coordinates is given by

$$x_2^2 = 4x_1^3 - Ax_1 - B,$$

for some $A, B \in k$, $He(E)$ is smooth if and only if $A \neq 0$ (cf. [Dol12, p. 139]). Then E is called *equianharmonic*. Now we want to define a map

$$\tau : He(E) \to He(E), \ a \mapsto \tau(a) = Sing(P_a(E)).$$

First, we will show that τ is well defined, i.e. we have $\tau(a) \in He(E)$ and $Sing(P_a(E))$ consists just of one point. Since we have $a \in He(E)$, we obtain by Lemma 2.2.2 that $P_a(E)$ is reducible, i.e. it is the union of two lines ℓ and ℓ'. Assume that $\ell = \ell'$. This implies that all points in $P_a(E)$ are singular, hence $Sing(P_a(E)) = \ell$. By Lemma 2.2.3, this means that $\ell \subseteq He(E)$. This is a contradiction to the smoothness of $He(E)$. Therefore $P_a(E)$ is the union of two distinct lines and hence $Sing(P_a(E))$ consists of one point. By Lemma 2.2.3, we conclude that this point indeed lies in $He(E)$. This proves that τ is well defined. It is called the *Steinerian automorphism*. The next step is to show the following

Lemma 2.3.4. *Let E be smooth plane cubic curve such that $He(E)$ is smooth. Then the map*

$$\tau : He(E) \to He(E), \ a \mapsto \tau(a) = Sing(P_a(E))$$

is well defined and a fixed point free involution, i.e. we have $\tau^2(a) = a$ and $\tau(a) \neq a$ for all $a \in He(E)$.

Proof. The first part was shown above, hence let us show the second part. First, we show that τ is an involution.
By Lemma 2.1.7, we know that

$$\tau(a) \in Sing(P_a(E)) \iff D_{\tau(a)}(D_a(F)) = 0.$$

By Lemma 2.1.2, this is equivalent to $D_a(D_{\tau(a)}(F)) = 0$, and again by Lemma 2.1.7 this holds if and only if $a \in Sing(P_{\tau(a)}(E))$. Since $a, \tau(a) \in He(E)$, which is smooth, we know that $Sing(P_a(E))$ resp. $Sing(P_{\tau(a)}(E))$ consist of $\tau(a)$ resp. a and thus $\tau^2 = id$. To show that τ has no fixed points, we assume that it has one, i.e. there exists an element $x \in He(E)$ such that $x = Sing(P_x(E))$. By Lemma 2.1.7 this holds if and only if $D_x(D_x(F)) = D_{x^2}(F) = 0$. By Lemma 2.1.6, this implies that $x \in Sing(E)$, a contradiction to the smoothness of E. $\square$

By means of τ we define the map

$$\eta : He(E) \to \mathbb{P}^2, \ a \mapsto \overline{a, \tau(a)},$$

which is well defined, since τ is fixed point free. This induces the map

$$\check{\eta} : He(E) \to \check{\mathbb{P}}^2, \ a \mapsto \overline{a, \tau(a)},$$

where the lines are considered as points in the dual plane $\check{\mathbb{P}}^2$ by means of the quadratic form $Z_0^2 + Z_1^2 + Z_2^2$. Now we define the *Cayleyan curve Ca(E)* of E to be the image of $He(E)$ under $\check{\eta}$, i.e. $Ca(E) := \mathrm{im}(\check{\eta})$.

We want to prove that $\check{\eta}$ induces a well-defined map

$$Ca : \mathcal{P}_{reg} \to \check{\mathcal{P}} := \{\text{Members of the dual Hesse pencil}\},$$

where $\mathcal{P}_{reg}$ denotes, as already used, the set of smooth members of the Hesse pencil. We will even show that the $Ca(E_\lambda)$ is the member of the dual Hesse pencil corresponding to the parameter $\frac{54-\lambda^3}{9\lambda}$. To see this, we need the following

Lemma 2.3.5. *Let E be a smooth cubic curve such that $He(E)$ is smooth. If a is a point of $He(E)$, then the second polar curve $P_{a^2}(E)$ is the tangent line to $He(E)$ at $\tau(a)$.*

Proof. We refer to Proposition 3.2 in [Kuw11]. $\qquad\qquad\square$

Our next goal is to connect the Steinerian automorphism with the group law of the elliptic curve. Therefore we fix again a smooth member E_λ of the Hesse pencil such that $He(E_\lambda)$ is smooth. Let us fix a group law on $He(E_\lambda)$ with $p_{i,j}$ as origin. We want to show that the Steinerian automorphism τ correspond to the translation by a non-trivial 2-torsion point. We will do this in several steps. The first step is the following

Lemma 2.3.6. *Let E_λ be a smooth member of the Hesse pencil such that $He(E_\lambda)$ is smooth. Let us fix a group law on $He(E_\lambda)$ by means of any inflection point $p_{i,j}$ as origin. Then $\tau(p_{i,j})$ is a non-trivial 2-torsion point.*

Proof. First, we remark by Lemma 2.3.4 that $\tau(p_{i,j}) \neq p_{i,j}$ and therefore it is non-trivial. We recall that $\tau(p_{i,j})$ is a 2-torsion point if and only if $p_{i,j} \in \mathbb{T}_{\tau(p_{i,j})}(He(E_\lambda))$. By Lemma 2.3.5, we have

$$\mathbb{T}_{\tau(p_{i,j})}(He(E_\lambda)) = P_{p_{i,j}^2}(E_\lambda).$$

Furthermore, we get by Lemma 2.1.3 that $p_{i,j} \in P_{p_{i,j}^2}(E_\lambda)$ if and only if $p_{i,j} \in E_\lambda$. We conclude that $\tau(p_{i,j})$ is a 2-torsion point if and only if $p_{i,j} \in \mathbb{T}_{\tau(p_{i,j})}(He(E_\lambda)) = P_{p_{i,j}^2}(E_\lambda)$ if and only if $p_{i,j} \in E_\lambda$. This completes the proof. $\qquad\qquad\square$

We now define $y := \tau(p_{i,j})$ and the translation map

$$t_y : He(E_\lambda) \to He(E_\lambda), \; x \mapsto x + y.$$

Clearly, this map is bijective. By Lemma 2.3.4, we have $\tau^2 = id$ and therefore τ is also bijective. We consider now the composition

$$t_y \circ \tau : He(E_\lambda) \to He(E_\lambda), \; x \mapsto \tau(x) + y.$$

By Lemma 2.3.6, we have that $y+y = 0$ and therefore $t_y \circ \tau(0) = y+y = 0$. This shows, that $t_y \circ \tau$ is an *isogeny*, i.e. $t_y \circ \tau(0) = 0$ (cf. [Sil86, Chap. III, §4]). Furthermore, by Theorem 4.8 in [Sil86, Chap. III, §4] we have that $t_y \circ \tau$ is a homomorphism of the elliptic curve $He(E_\lambda)$. Since t_y and τ are bijective, we conclude that $t_y \circ \tau$ is also bijective and hence is an automorphism of $He(E_\lambda)$. The automorphism group $Aut(He(E_\lambda))$ is classified in Theorem 10.1 in [Sil86, Chap. III, §10]. By this, the set of elliptic curves E_λ with $|Aut(E_\lambda)| = 2$ is dense in the Hesse pencil. Therefore by a zariski closure argument, it suffices to consider the elliptic curves E_λ with $|Aut(E_\lambda)| = 2$. Since the group defined in 1.4.1 is abelian, we have the identity and the map $x \mapsto -x$ in $Aut(He(E_\lambda))$. Therefore it suffices to show that for some point $0 \neq x \in He(E_\lambda) \smallsetminus He(E_\lambda)[2]$ holds $t_y \circ \tau(x) = x$, where $He(E_\lambda)[2]$ denotes, as usual, the group of 2-torsion points. We will do it by a straightforward computation. It suffices to show that for any $p_{i,j}$ as origin of the group law exists another inflection point $p_{k,l}$ such that $t_y \circ \tau(p_{k,l}) = p_{k,l}$. These computations are essentially symmetrical, but nevertheless there are nine cases to check. We will do one case in full detail and leave the other cases to the reader.

Fix a group law on $He(E_\lambda)$ by means of $p_{0,0} = [0 : 1 : -1]$ as the origin. By a computation we have $\mathbb{T}_{p_{0,0}}(E_\lambda) = V(-\lambda Z_0 + 3Z_1 + 3Z_2)$ and $L_{0,0} = V(Z_1 - Z_2)$. Therefore we get

$$\tau(p_{0,0}) = V(-\lambda Z_0 + 3Z_1 + 3Z_2) \cap V(Z_1 - Z_2) = [\frac{6}{\lambda} : 1 : 1].$$

Note that we have $\lambda \neq 0, 6$, since $He(E_\lambda)$ was assumed to be smooth. Then we chose $p_{1,0} = [1 : 0 : -1]$ and compute analogously

$$\tau(p_{1,0}) = V(3Z_0 - \lambda Z_1 + 3Z_2) \cap V(Z_0 - Z_2) = [1 : \frac{6}{\lambda} : 1].$$

Now we have to show

$$t_{\tau(p_{0,0})} \circ \tau(p_{1,0}) = \tau(p_{1,0}) + \tau(p_{0,0}) = p_{1,0},$$

i.e. that the third common point of $\overline{p_{1,0}, p_{0,0}}$ and $He(E_\lambda)$ coincides with the third common point of $\overline{\tau(p_{1,0}), \tau(p_{0,0})}$ and $He(E_\lambda)$. We note that $\overline{p_{1,0}, p_{0,0}}$ resp. $\overline{\tau(p_{1,0}), \tau(p_{0,0})}$ are given by

$$Z_0 + Z_1 + Z_2 = 0$$

resp.

$$Z_0 + Z_1 - (1 + \frac{6}{\lambda})Z_2 = 0.$$

These two lines intersect in $p_{2,0} = [-1 : 1 : 0]$, which also lies in $He(E_\lambda)$, since it is a basepoint of the Hesse pencil by Lemma 1.1.10. This shows that $\tau(p_{1,0}) + \tau(p_{0,0}) = p_{1,0}$ and therefore $t_{\tau(p_{0,0})} \circ \tau$ has to be the identity on $He(E_\lambda)$. The other cases are similar. We conclude:

Lemma 2.3.7. *Let E_λ be a smooth member of the Hesse pencil such that $He(E_\lambda)$ is smooth. Fix any group law on $He(E_\lambda)$. Then, for some $0 \neq y \in He(E_\lambda)[2]$, the Steinerian automorphism τ is given by $\tau(x) = x + y$, for all $x \in He(E_\lambda)$.*

The second part of the next lemma is known by the properties of the Hesse pencil, but we want to give an example of the use of the technique induced by Lemma 2.3.7.

Lemma 2.3.8. *Let E_λ be a smooth member of the Hesse pencil such that $He(E_\lambda)$ is smooth, and let a be an inflection point of E_λ. Then holds:*

> *i) The inflection tangent line $\mathbb{T}_a(E_\lambda)$ is again tangent to $He(E_\lambda)$ at $\tau(a)$. In particular, $\mathbb{T}_a(E_\lambda)$ coincides with the line $\overline{a, \tau(a)}$;*
>
> *ii) a is also an inflection point of $He(E_\lambda)$.*

Proof. We follow the proof of Proposition 3.3 in [Kuw11]. To prove *i)* we have by Lemma 2.2.4 that $\mathbb{T}_a(E_\lambda)$ coincides with the polar second polar $P_{a^2}(E_\lambda)$. By Lemma 2.3.5, we get

$$P_{a^2}(E_\lambda) = \mathbb{T}_{\tau(a)}(He(E_\lambda))$$

and the assertion holds.

To see *ii)* choose one of the inflection points of $He(E_\lambda)$ as origin of the group law. By *i)* the line $\overline{a, \tau(a)}$ is tangent to $He(E_\lambda)$ at $\tau(a)$. Therefore we get $a + \tau(a) = -\tau(a)$. If we use now that $\tau(a) = a + y$, for some $0 \neq y \in He(E_\lambda)[2]$, we get $a + 2\tau(a) = a + 2(a + y) = 3a + 2y = 3a$. Therefore we have $3a = 0$ and thus a has to be an inflection point of $He(E_\lambda)$. □

Now by Lemma 2.3.5 and Lemma 2.3.7, we are able to give a very important description of the Cayleyan by means of the next theorem. In its proof, we will use the definition of the group law several times without further comments.

Theorem 2.3.9. *Let E_λ be a smooth member of the Hesse pencil such that $He(E_\lambda)$ is smooth. Then a line $\ell \in \check{\mathbb{P}}^2$ belongs to $Ca(E_\lambda)$ if and only if it is an irreducible component of the first polar curve $P_d(E_\lambda)$ for any $d \in He(E_\lambda)$.*

Proof. We fill out the details of the proof of Proposition 3.4 in [Kuw11]. Let $\ell = \overline{a, \tau(a)} \in Ca(E_\lambda)$, for some $a \in He(E_\lambda)$. Let us fix some group law on $He(E_\lambda)$. We have by Lemma 2.3.7 that $-2\tau(a) = -2a$. By Definition 1.4.1, we get that $-2a$ is the third point in $\mathbb{T}_a(He(E_\lambda)) \cap He(E_\lambda)$, resp. $-2\tau(a)$ is the third point in $\mathbb{T}_{\tau(a)}(He(E_\lambda)) \cap He(E_\lambda)$. Since $-2\tau(a) = -2a$, we have

$$-2a \in \mathbb{T}_a(He(E_\lambda)) \cap \mathbb{T}_{\tau(a)}(He(E_\lambda)) \cap He(E_\lambda).$$

29

By Lemma 2.2.4, we have

$$\mathbb{T}_a(He(E_\lambda)) = P_{a^2}(He(E_\lambda))$$

resp.

$$\mathbb{T}_{\tau(a)}(He(E_\lambda)) = P_{\tau(a)^2}(He(E_\lambda))$$

and thus

$$-2a \in P_{\tau(a)^2}(He(E_\lambda)) \cap P_{a^2}(He(E_\lambda)) \cap He(E_\lambda).$$

We will show that $d := -2a$ is the point such that $\overline{a, \tau(a)} \subseteq P_d(E_\lambda)$. Since $d \in He(E_\lambda)$ and $He(E_\lambda)$ is smooth, we know that $P_d(E_\lambda)$ is the union of two distinct lines ℓ and ℓ'. Furthermore we know that the intersection of ℓ and ℓ' is given by $d + y = -2a + y$ for some $0 \neq y \in He(E_\lambda)[2]$. We have

$$-a - \tau(a) = -a - (a + y) = -2a - y = -2a + y = d + y$$

and hence $d + y \in \overline{a, \tau(a)} \cap He(E_\lambda)$.

Now we claim that $a \in P_d(E_\lambda)$. We have $\tau(a) + \tau(a) = -d$ and hence $d \in \mathbb{T}_{\tau(a)}(He(E_\lambda))$. Furthermore, we have by Lemma 2.3.5 that $\mathbb{T}_{\tau(a)}(He(E_\lambda)) = P_{a^2}(E_\lambda)$ and therefore $d \in P_{a^2}(E_\lambda)$. By Lemma 2.1.3, this holds if and only if $a \in P_d(E_\lambda)$ and the claim is proved.

We conclude that $\overline{a, \tau(d)} \subseteq P_d(E_\lambda)$ and thus the line coincides with an irreducible component of $P_d(E_\lambda)$. Since $\tau(d) = d + y \in \overline{a, \tau(a)} \cap He(E_\lambda)$, we obtain that the line $\overline{a, \tau(d)}$ coincides with the line $\overline{a, \tau(a)}$ and hence $\overline{a, \tau(a)} \subseteq P_d(E_\lambda)$ and the line ℓ is an irreducible component of a first polar of E_λ.

Conversely, consider $P_d(E_\lambda)$ for some $d \in He(E_\lambda)$. Therefore $P_d(E_\lambda)$ is the union of two lines (cf. Lemma 2.2.2). Since $He(E_\lambda)$ is smooth, it is even the union of two distinct lines intersecting at $\tau(d)$. We consider the equation $2x = -d$. It certainly has one solution a. Furthermore, for any additional nontrivial 2-torsion points y' we get the four solutions $a, a+y, a+y'$ and $a+y+y'$. Let us define $a' := a+y'$. Now, the argument in the end of the first part of the proof shows that the lines $\overline{a, \tau(a)}$ and $\overline{a', \tau(a')}$ coincide with the irreducible components of $P_d(E_\lambda)$ and thus the proof is completed. $\square$

This exact description of the Cayleyan curve will help us to prove our last

Theorem 2.3.10. *Let E_λ be a member of the Hesse pencil. Then its Cayleyan curve $Ca(E_\lambda)$ is the member of the dual Hesse pencil corresponding to the parameter $\frac{54-\lambda^3}{9\lambda}$.*

Proof. We follow the arguments of Proposition 3.3 in [AD10]. We use the characterization of the Cayleyan of Theorem 2.3.9. For aesthetic reasons, we fix $E_{6\mu}$ and consider the first polar $P_q(E_{6\mu})$ of the point $q = [u : v : w] \in \mathbb{P}^2$ with respect to $E_{6\mu}$. It is given by

$$u(Z_0^2 + 2\mu Z_1 Z_2) + v(Z_1^2 + 2\mu Z_0 Z_2) + w(Z_2^2 + 2\mu Z_0 Z_1) = 0.$$

This polar is reducible if the equation decomposes into linear factor, i.e.

$$u(Z_0^2 + 2\mu Z_1 Z_2) + v(Z_1^2 + 2\mu Z_0 Z_2) + w(Z_2^2 + 2\mu Z_0 Z_1)$$

$$= (aZ_0 + bZ_1 + cZ_2)(\alpha Z_0 + \beta Z_1 + \gamma Z_2),$$

for some $a, b, c, \alpha, \beta, \gamma \in k$. By a comparison of coefficients, this happens if and only if

$$\begin{pmatrix} u & 2\mu w & 2\mu v \\ 2\mu w & v & 2\mu u \\ 2\mu v & 2\mu u & w \end{pmatrix} = \begin{pmatrix} a\alpha & a\beta + b\alpha & a\gamma + c\alpha \\ a\beta + b\alpha & b\beta & c\beta + b\gamma \\ a\gamma + c\alpha & c\beta + b\gamma & c\gamma \end{pmatrix}.$$

Now, we consider this as a system of linear equations in the variables u, v, w, a, b and c. Then it has a non-trivial solution if and only if the following determinant vanishes:

$$\begin{vmatrix} -1 & 0 & 0 & \alpha & 0 & 0 \\ 0 & -1 & 0 & 0 & \beta & 0 \\ 0 & 0 & -1 & 0 & 0 & \gamma \\ -2\mu & 0 & 0 & 0 & \gamma & \beta \\ 0 & -2\mu & 0 & \gamma & 0 & \alpha \\ 0 & 0 & -2\mu & \beta & \alpha & 0 \end{vmatrix} = \mu(\alpha^3 + \beta^3 + \gamma^3) + (1 - 4\mu^3)\alpha\beta\gamma = 0.$$

Note that this system of linear equations only consists of six equations, since the matrices given above are symmetrical. Now we take $[\alpha : \beta : \gamma]$ as the coordinates in the dual plane $\check{\mathbb{P}}^2$. The line $\alpha Z_0 + \beta Z_1 + \gamma Z_2 = 0$ is an irreducible component of the first polar $P_q(E_{6\mu})$ and hence it characterizes by Theorem 2.3.9 the Cayleyan curve $Ca(E_{6\mu})$. By setting $\mu := \frac{\lambda}{6}$, the proof is completed. $\square$

References

[AD10]   ARTEBANI, Michela ; DOLGACHEV, Igor V.: The Hesse pencil of
         plane cubic curves. In: *L'Enseignement Mathématique* 55 (2010),
         S. 1–39

[Bea96] BEAUVILLE, Arnaud: *Complex Algebraic Surfaces -*. 2nd ed. Cam-
         bridge : Cambridge University Press, 1996

[BH85]   BARTH, Wolf ; HULEK, Klaus: Projective models of Shioda mod-
         ular surfaces. In: *Manuscripta Mathematica* 50 (1985), S. 73–132

[BK86]   BRIESKORN, Egbert ; KNÖRRER, Horst: *Plane algebraic curves*.
         1st ed. Basel : Birkhäuser Verlag, 1986

[Dol12] DOLGACHEV, Igor V.: *Classical Algebraic Geometry - A Modern
         View*. 1st ed. Cambridge : Cambridge University Press, 2012

[Ful08] FULTON, William: *Algebraic curves - An Introduction to Algebraic
         Geometry*.   www.math.lsa.umich.edu/~wfulton/CurveBook.pdf,
         2008

[Kuw11] KUWATA, Masato: Constructing families of elliptic curves with
         prescribed mod 3 representation via Hessian and Cayleyan curves.
         In: *Preprint* (2011)

[PP09]   POPESCU-PAMPU, Patrick: Iterating the hessian: a dynamical
         system on the moduli space of elliptic curves and dessins d'enfants.
         In: *Adv. Stud. in Pure Math.* 55 (2009), S. 83–98

[Sil86] SILVERMAN, Joseph H.: *The Arithmetic of Elliptic Curves*. 1st ed.
         Berlin, Heidelberg : Springer, 1986

YOUR KNOWLEDGE HAS VALUE

- We will publish your bachelor's and master's thesis, essays and papers

- Your own eBook and book - sold worldwide in all relevant shops

- Earn money with each sale

Upload your text at www.GRIN.com and publish for free